ADVANCE PRAISE

"This book, written by distinguished physician Dr Don Simborg, can be read in many ways and on many levels. As a reader outside of this field, and indeed not a philosopher, I found the book fascinating. While reading this carefully, scientifically well-documented and referenced document and questioning again and again: 'What is life?', what stood out to me were the detailed discussions referencing so many distinguished experts who, for centuries, have addressed the same question.

"It made me think of how, for over 5,000 years, Talmudic scholars have questioned, discussed, argued and documented their personal philosophical discussions on the meaning of God. Looking up how Talmudic Judaism Jews seek truth, it was defined as follows: 'Classical rabbinic study is organized around arguments, questions and counter-arguments rather than linear exposition.' This seems to me a very similar approach presented to us by Dr Simborg in the book, offering a careful assessment in search of an answer.

"I will leave it to the reader to enjoy and see what Dr Simborg concludes. Instead of following a religious approach, he uses a more scientific and philosophical approach to address the fascinating question: 'What is life?'"

MARION J BALL, EdD, Presidential Distinguished Professor, University of Texas at Arlington

"What is Life? *is Don Simborg's invitation to succinctly and boldly explore the depth and breadth of tantalizing questions, ranging from vitalism, DNA/RNA, atomism, negative entropy, order from order, wave theory, pillars of life, and extraterrestrial life. This artful reading begs discernment of micro and macro perspectives; the question 'what is life' becomes more, not less, complex. The mind-heart connection surges with the thirst for simplicity. As T. S. Eliot reminds us "We shall not cease from exploration, and the end of all our exploring will be to arrive where we started and know the place for the first time." And Don gently brings us back to the start and to perhaps the most life-giving definition noted in his early pages about Homo sapiens 3,000+ years ago… soul, vitality, essence and later coupled with purpose-driven existence, discovery and transformation that are most respected, honored and understood from the collective, holistic, interconnected, non-reducible infinity."*

CONNIE WHITE DELANEY, PhD, RN, Professor and Dean,
University of Minnesota School of Nursing

"In the past we used to ask ourselves what distinguishes life from dead material – today we are wondering about the difference between (human) life and artificial intelligence in the future. By putting a question from the past this book equips you with thoughts for debating tomorrow's fundamental question."

URSULA H. HÜBNER, Phd, FIAHSI, School of Business
Management and Social Sciences, Osnabrück University
Applied Sciences

"Simborg asks the eternal question, 'What is life'?' in a thoughtful, perceptive, educational, and entertaining way.

Simborg's brief book is not religious or philosophical as much as it is scientific and educational, guiding the reader through a detailed journey that helps one to understand why the definitional task is difficult, while leading to insight coupled with a sense of wonder.

Without emphasizing the point, Simborg shoots holes in most science fiction stories that involve life in other parts of the universe.

Although the presentation involves significant scientific detail, revealing the author's breadth of knowledge and the detailed work he performed to study the topic, the volume does not overwhelm the uninitiated and instead draws the reader into the exploration as they learn that the question, and proposed answers, are neither new nor ultimately sufficient.

The book starts by asking whether a definition of life is even possible and then takes the reader on a journey through the many attempts, while explaining why a question that at first seems obvious is ultimately unanswerable. The cogent analysis leaves the reader with a sense of wonder regarding the complexity of the question and the insights gained as we learn why brilliant and knowledgeable people have been stymied in their efforts to be precise."

EDWARD H. SHORTLIFFE, MD, PhD, Chair Emeritus and Adjunct Professor, Department of Biomedical Informatics, Columbia University

Published by
LID Publishing
An imprint of LID Business Media Ltd.
LABS House, 15-19 Bloomsbury Way,
London, WC1A 2TH, UK

info@lidpublishing.com
www.lidpublishing.com

A member of:

BPR

businesspublishersroundtable.com

Reprinted in 2026

Printed and bound in Great Britain by Halstan

ISBN: 978-1-918215-35-9
ISBN: 978-1-918215-36-6 (ebook)

Cover and page design: Caroline Li
Illustrations created with the help of AI

WHAT IS LIFE?

THE QUEST FOR A DEFINITION

DON SIMBORG

LID

MADRID | MEXICO CITY | LONDON
BUENOS AIRES | BOGOTA | SHANGHAI

What is [my] life?

Maddy

CONTENTS

PREFACE

NON-LIFE

Yes, it's a fact. There is no agreed-upon scientific definition of life. It's not that we haven't tried. There have been attempts to define life for millennia. But each one seems to have a flaw or exception that makes the definition invalid or unacceptable or incomplete or simply controversial among scientists.

To many people, this seems difficult to understand. The average person would claim to know life when they see it. It seems so obvious. Plants and animals are alive and everything else isn't.

Well, that's not very scientific. A scientist would quickly point out that there are living things that are neither plants nor animals, like bacteria or fungi. OK, so our definition obviously needs to include those two categories as well. But, oops, that leaves out archaea, which most people have never heard of but are one of the most numerous forms of life. They look like bacteria – consisting of a single cell, visible only under a microscope – and are virtually everywhere.

You get the idea. Simply making a list of categories of things that most people consider to be alive may not be the best way to go about the process of defining life. There are a few problems with that approach. First, some categories of things that some people would consider to be alive others may not think are. For example, viruses: are they alive? There's a lot of debate about that. We'll get to this topic – and others that are similarly controversial – later in the book.

The second problem is what to include in any given category. There are many living things that, depending on how each category is defined, could arguably be in one or another or none of the above. Finally, as will be discussed in *Chapter 9*, synthetic biologists are creating new forms of life that don't easily fit into existing categories. So, simply making a list of these categories doesn't get to the heart of the definition of life. There must be a better way.

Rather than making lists of types of living things, scientists have attempted to define life as consisting of certain features, functions or characteristics. Are there one or two features or functions that all living things have and all non-living things don't have? For example, do all things that die define things that are alive? That depends on what we mean by 'die,' which is another complicated question. Doctors still debate what constitutes death in humans, and those debates don't even touch on all the other forms of life.

Nonetheless, there is no debate that humans die. But so do fire, the wind, governments, soundwaves and... OK, that's just semantics. But death implies that something essential for life stops, and maybe if we could define what that something is, it would constitute a definition of life. Breathing doesn't work since a lot of living things don't breathe (depending on how you define 'breathing'). What about DNA? All living things that we know of on Earth use DNA as their genetic molecule. Well, DNA can exist in a test tube solution, so that doesn't work. There are other problems with DNA too, as we'll see in later chapters.

So, what about cells? All living things consist of a single cell or multiple cells, where 'cell' is defined as a closed environment, surrounded by a lipid membrane, where chemical processes take place (scientific terms such as 'lipid membrane' are explained in the Glossary at the end of the book). This comes pretty close to fitting my criteria that all living things have the characteristic in question and all non-living things do not.

Close, but not quite. There are things called liposomes, which are vesicles (fluid-filled sacs) with lipid membranes that enclose any number of drugs and/or chemical reactions that can be created in a laboratory. Such vesicles do not contain any genetic material, do not have a life-sustaining metabolism and do not reproduce. They are clearly not alive. There are other similar structures within cells with scientific names that would require some length to describe here, but suffice it to say, we would need more complicated definitions to distinguish them from non-living things.

This brings us back to the most common attempts to define life. Rather than attempting to find a single or perhaps a couple of clear defining characteristics, is there a longer list of things that would work? Procreating? Metabolizing? Evolving? Having cells? Creating new cells? Using DNA and/or RNA? These are just some of the characteristics that people have used when making lists to define life.

We're now getting into the fundamental debates that have gone on for millennia. It seems that one or more of these characteristics, perhaps alongside others, should solve the problem.

It may just be a matter of finding which ones.

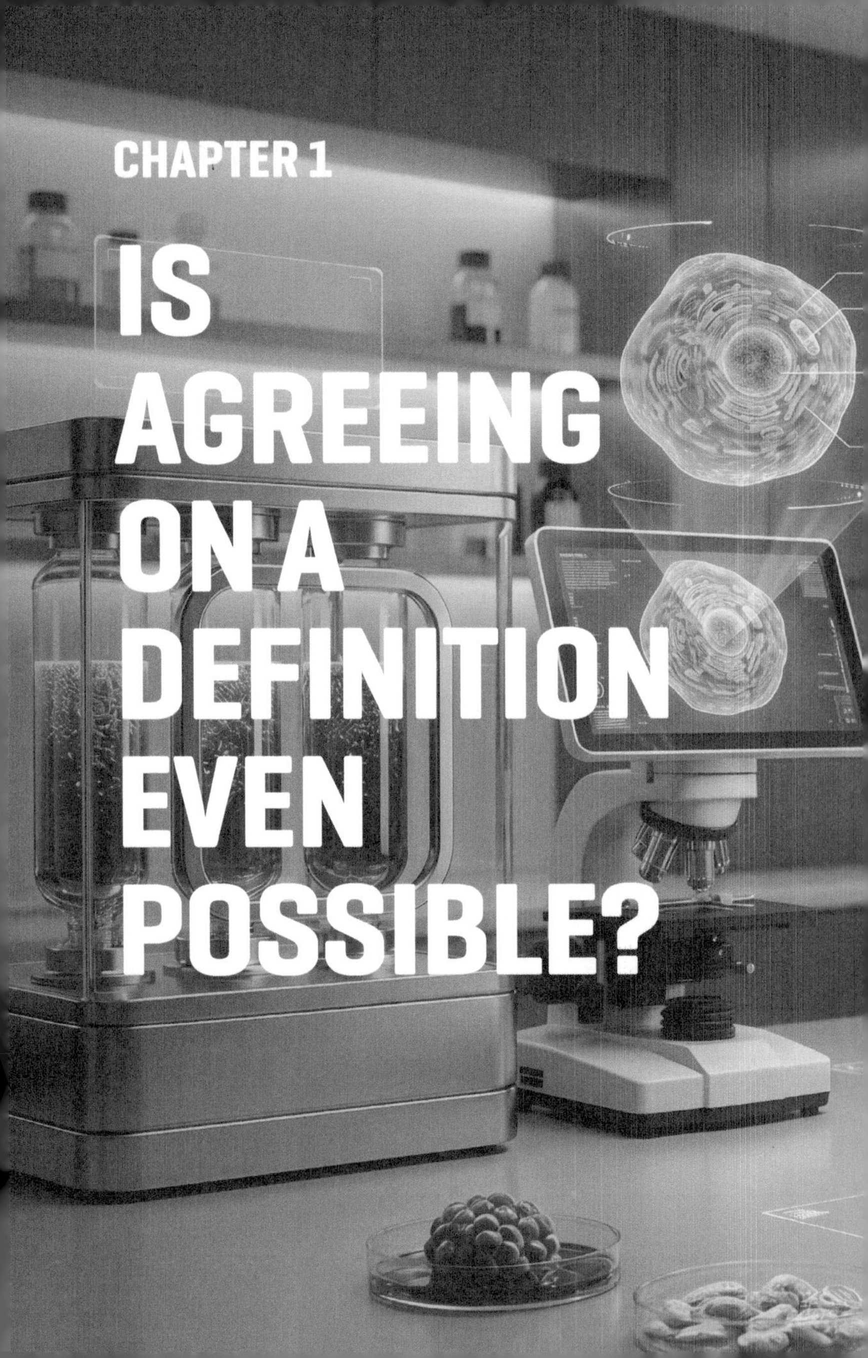

CHAPTER 1

IS AGREEING ON A DEFINITION EVEN POSSIBLE?

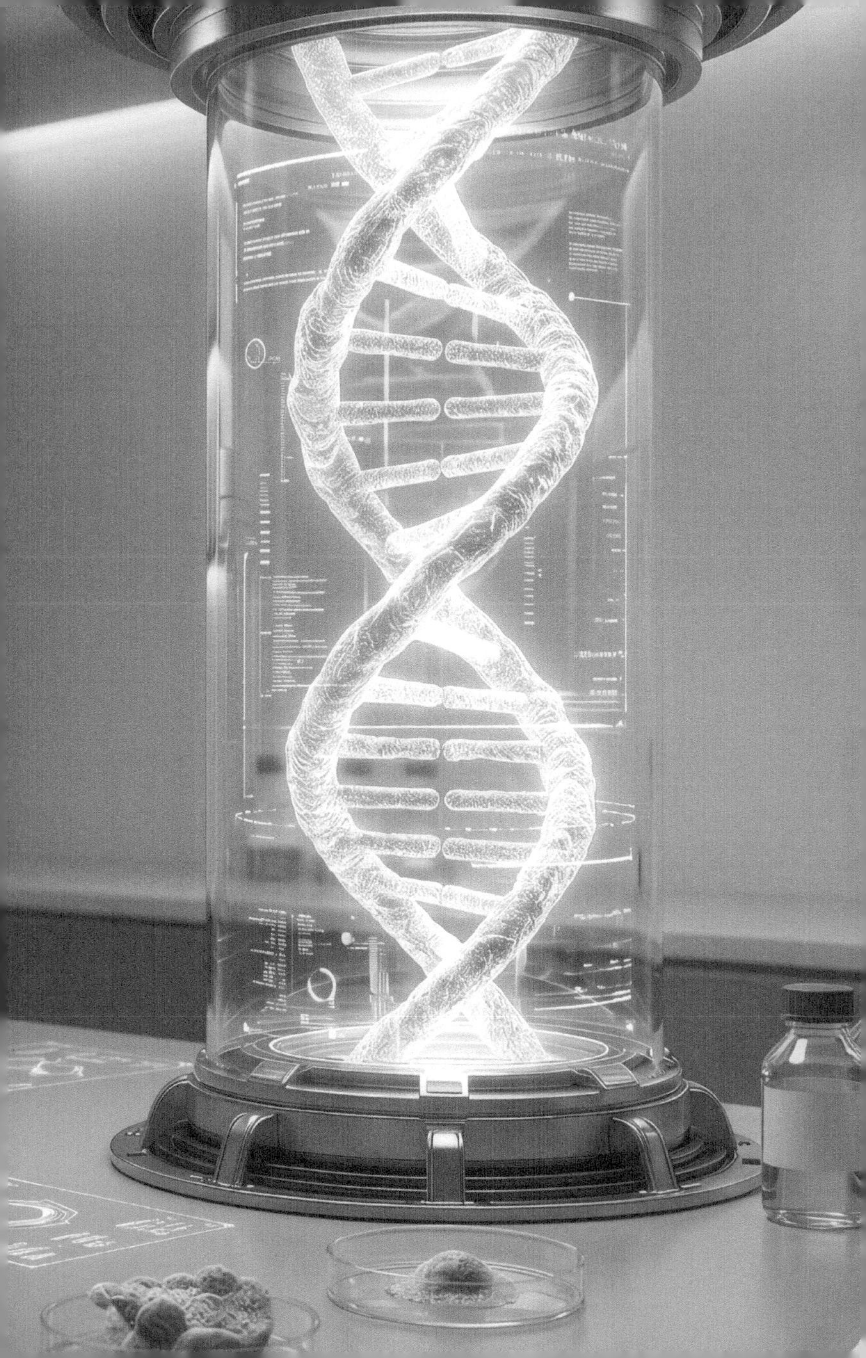

"Life is the essence that God imbued in some things when he or she created the universe." Some would consider this statement to be a definition of life. Others would say that it is just a belief and not a definition. We could argue that if it constitutes a definition, we will need to better characterize what that 'essence' is. Maybe it is the same thing as those characteristics we are trying to define. Or maybe we'd just end up with the circular conclusion that the 'essence' is only something God understands.

In any case, if we allow a religious or even a non-religious belief system to be involved in the definition, there is no need for further debate. There is no way to resolve such a discussion. Since the earliest attempts to define life, there has always been what I would call an 'objective' or 'scientific' thread as well as a 'non-objective' or 'non-scientific' thread. Although I will track both threads, this conflict cannot be resolved. The non-scientific thread, involving religious or other belief systems, will always exist. So, I will focus on trying to resolve the scientific thread.

CHAPTER 2

WHAT DOES *THE DICTIONARY* SAY?

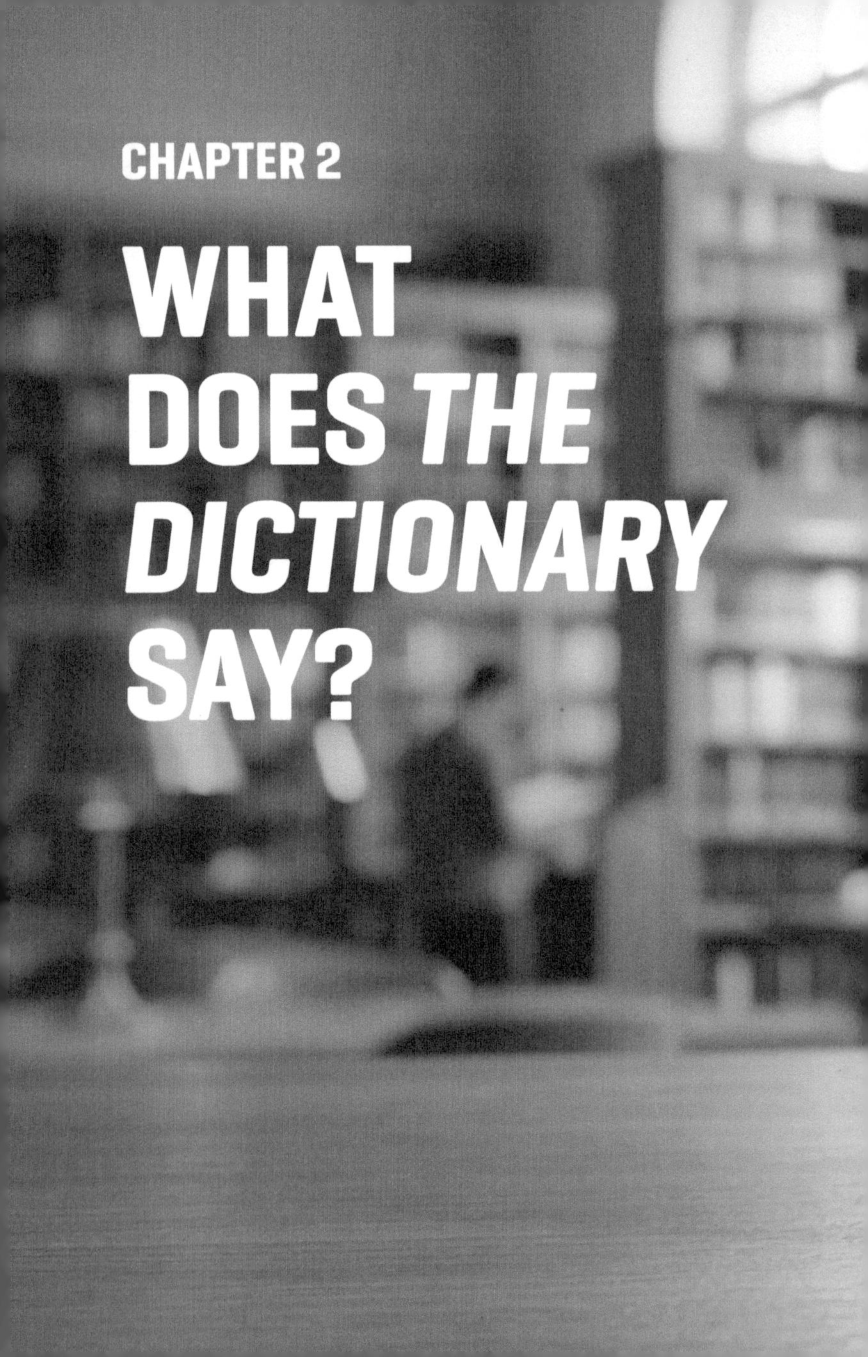

Merriam-Webster
Dictionary
Merriam-Webster
Dictionary

The primary definition of 'life' in the online *Merriam-Webster* dictionary is:[1]

> **a:** the quality that distinguishes a vital and functional being from a dead body
> **b:** a principle or force that is considered to underlie the distinctive quality of animate beings
> **c:** an organismic state characterized by capacity for metabolism, growth, reaction to stimuli, and reproduction[1]

This definition doesn't help us much as it reflects the age-old conflict discussed in the preface. Regarding the first definition, we've already seen that we need to define the quality that distinguishes a living from a dead body – not just state that there is one. The second definition suffers from the same problem. What exactly is that 'principle' or 'force'? The third comes closer to what we're seeking: it's a list of characteristics that might be used to define life. What could be wrong with life being defined as the capacity for metabolism, growth, reaction to stimuli and reproduction?

Well, a mule can't reproduce. Nor can some humans after a certain age. So, does that mean a mule and those humans are not alive? And some organisms can enter states where they no longer react to stimuli.

When it comes to growth, we all stop growing at some point (if, by 'growth,' we mean a net increase in our volume or number of cells). Bacteria don't grow. In contrast, lots of non-living things grow, such as crystals, the universe and glaciers. And what about those pesky viruses? They don't grow. Are we assuming they are not alive?

Finally, what do we mean by metabolism? This is just a set of chemical reactions, which can be found in non-living environments as well. And some living organisms in certain frozen or inert states don't metabolize.

Fire grows, responds to stimuli and causes chemical reactions akin to metabolism. Is fire alive?

But the third *Merriam-Webster* definition simply states that living things have the 'capacity' to do those four things, not that they necessarily do them all at the same time. Is that the clue we're looking for? Is life the 'capacity' to do certain things? And, if we can find the right combination of things, can the capacity to do them all at some time define life and the lack of capacity to do them all define non-life?

Another comment about the third *Merriam-Webster* definition is that it leaves out certain things that most people believe are necessary to define life, such as the ability to evolve by responding to the environment through natural selection.

Is defining life just a matter of getting agreement on the full list of requirements?

CHAPTER 3

SOME FACTS OF LIFE

Life on Earth emerged 3.8 billion years ago.* Let's call that a fact even though some might quibble about the exact timing. But that fact is surrounded by a whole host of questions. Did life spontaneously emerge on Earth, or did it arrive here on one of the many celestial objects that we know were bombarding the surface of the planet at that time? If it did emerge spontaneously on Earth, how exactly did that happen? From what chemicals and what conditions? We certainly have never been able to re-create life, despite scientists having experimented with a whole host of chemicals and conditions.

Another interesting fact is that if life emerged spontaneously on Earth 3.8 billion years ago, it has not done so again since. That is, there is no known lineage of life on Earth that dates to a later time. Thus, the conditions that created life on Earth have not recurred or they never were present. Life creation – as well as life itself – is a unique phenomenon that currently defies definition.

Nevertheless, however and wherever life began, it is a fact that life exists on Earth. Further, the forms that life takes have clearly evolved. The variety of life forms has increased dramatically over the past 3.8 billion years. This has *not* happened because of new forms spontaneously emerging, but rather because of evolution from previous forms, all dating back to some original life form. It is still unknown whether

* The Earth formed 4.5 billion years ago. It took a while for it to cool sufficiently and for other conditions to develop to enable life to emerge and be sustained.

that original life form was a single-celled organism (such as an archaeon or bacterium), some sort of RNA-based or virus-like entity, or something else. Thus, the question of how life was created, just like the question of how to define life, remains unresolved. As will be discussed in *Chapter 7*, attempts to define life are often closely linked to attempts to understand how life was originally created.

Another important point to consider here is that there is a difference between defining life and defining living things. The only life we *Homo sapiens* have experienced is here on Earth. Life could be quite different in its fundamentals elsewhere in the universe. Maybe life elsewhere does not require DNA or RNA, or water, or carbon, or some other fundamental characteristic that many current definitions of life presume. For example, there has been speculation that perhaps methane or ammonia could replace water as the life-sustaining solvent, or silicon could replace carbon, or any number of other complex molecules could replace DNA and RNA.[2]

No evidence for any of these substitutions has been found in living things in nature, but we haven't explored much of the universe yet. And, in the little we have explored so far, our search for extraterrestrial life has mostly been focused on places in our own solar system where water exists or has existed. So perhaps we should seek a definition of life at a higher or more abstract level that avoids Earth-centric fundamentals.

CHAPTER 4

THE EARLIEST ATTEMPTS AT A DEFINITION

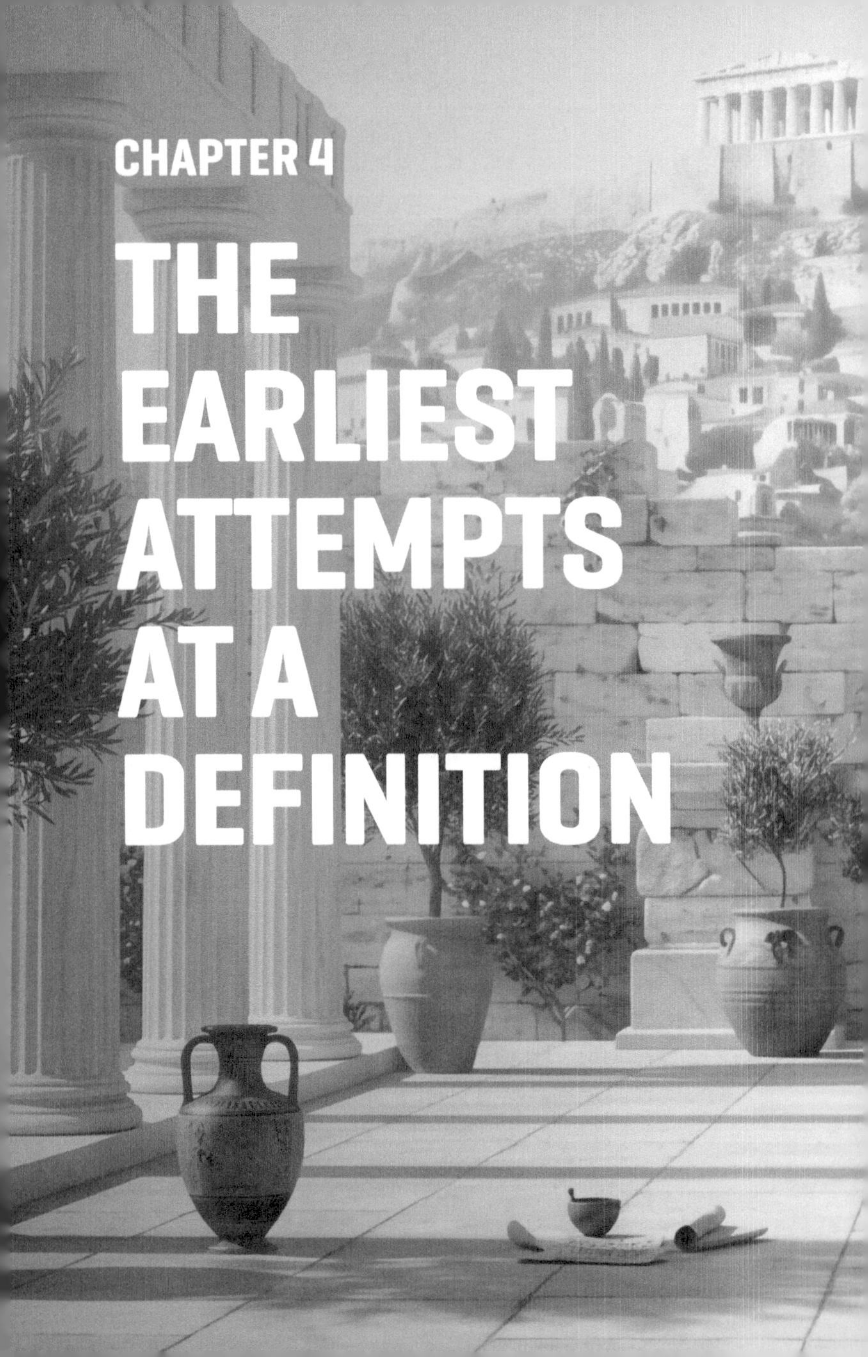

I'm going to take it as a given that non-humans do not have the capacity to consider and communicate a definition of life. Therefore, the earliest known attempts must have come from a human species, the earliest of which appeared a couple of million years ago. However, since even human species had no capability to communicate complex thoughts through language (recorded or not) for most of that period, the earliest possible attempts to define life must have happened no more than 100,000 years or so ago. Only during that period could verbal thoughts about life have been passed on to future generations.

When it comes to the earliest *recorded* attempts to define life, these date from only around three millennia ago. By that point, earlier species of humans – *Homo erectus*, Neanderthals, Denisovans and all others – had long died out. Thus, the question of this book is uniquely related to *Homo sapiens* (us). What is life as defined by *Homo sapiens* and, at least for the time being, as observed on Earth?

The ancient Greeks have been credited with the earliest discussions of the nature of life. Democritus, who lived in the 5th century BCE, was one of the earliest Greek philosophers to espouse a theory of life.[3] He was a believer in atomism, which states that all things, living and non-living, consist of tiny, too-small-to-be-visible-to-the-naked-eye particles. These particles, or atoms, vary in size, shape and quality in a way that, when the particles are aggregated, determines the physical and functional characteristics of an object.

The only thing that is 'non-atomic' is simply a void through which atoms can move. In the case of living things, these atoms are 'fiery,' which accounts for the heat associated with living things (and seems to preclude plants). Death is simply the dissipation of these fiery atoms, and there is no 'soul' or afterlife.

In atomism, then, life is the aggregation of fiery atoms. Although the idea is simple, and not totally inconsistent with some modern definitions (doesn't everything consist of atoms?), it leaves a lot of unanswered questions.

About the same time as Democritus, another Greek philosopher, Empedocles, had a more complex view of life that had many ingredients reflected in today's science and religion.[4] Empedocles claimed that all living things have the necessary mixture of the basic components of the universe: fire, water, air and earth. In addition, he believed that two vital forces determine how those components mix and work together: love and strife. Things live when love dominates and die when strife dominates.

Upon death, something beyond the four components, a soul, persists. It is reunited with a cosmic soul and later reincarnated into future living things. Empedocles even had a pre-Darwinian idea that those living things best able to survive and reproduce in their environment are those that will persist over generations.

So, as early as 3,000 years ago, the great *Homo sapiens* thinkers had introduced atomic theory, the concept of fundamental forces (we now have four rather than two)* and evolutionary theory. These are key scientific concepts today.† They had also established the beginnings of our religious thinking regarding a soul or some other non-material 'essence' or 'vitality' that imbues life in a mysterious or God-like fashion.

Every definition of life since can be considered a variation on these two threads (scientific vs. non-scientific). Both threads began with the earliest attempts of *Homo sapiens* to define life. Both threads remain even as the human brain and the science it has created have evolved.

---

* The four forces of nature are electromagnetism, the strong nuclear force, the weak nuclear force and gravity.

† Some argue today that there are only two fundamental scientific concepts: physics (which would include atomic theory and fundamental forces) and evolutionary science (called the "second science" in Mark Vellend's book *Everything Evolves*).

CHAPTER 5

VITALISM AND FORCES

René Descartes, French philosopher, 1596 - 1650

The concept of a soul or 'vital' force is generally discounted today by scientists seeking to define life. In Plato's *Timaeus*, a universal soul, infused by the creator,[5] permeates all things, including the planets, giving them life. In addition, humans and other animals have an individual soul that infuses life into their physical bodies. This soul consists of the four universal elements: fire, water, air and earth (same as Empedocles). Thus, 'life' and 'soul' seemed to be synonymous. Moreover, it is clear that in Plato's mind, entities that no one today would consider 'living' had a soul.

Interestingly, Plato also thought that life (or the soul) brought order to the universe, which was otherwise disordered. This idea resonates with today's discussion of entropy and the role of life, which temporarily violates the second law of thermodynamics (which states that in all natural systems, disorder (or entropy) increases with time, as discussed in the next chapter).

Plato's student Aristotle had a different view of the soul. He said that only living things possess a soul, which distinguishes them from the non-living physical world.[6] The soul enables living things to maintain 'order,' allowing resistance to external disturbances and promoting growth and reproduction. This concept – that some kind of vital non-material force conveys 'life' – has persisted throughout recorded human history. It is often referred to as 'vitalism.'

Belief in vitalism peaked in the 18th and 19th centuries, and the idea was widely accepted as fact by scientists and medical professionals. The inability to fully explain life processes through confirmed scientific and objective measures fuelled this belief. Even renowned scientists such as Louis Pasteur believed that some unknown (and unknowable) 'vital action' distinguished living from non-living things and enabled growth, reproduction and other functions found only in the living.[7]

In contrast, one of the greatest philosophers of the 17th century, René Descartes,[8] rebuked vitalism by claiming that all living organisms, including the human body, can be explained entirely by physical laws and mechanical principles. There was no need to invoke any vital force to explain life. Yet even Descartes proposed the notion of an immaterial and unknowable soul that exists only in humans and that accounts for the human mind and consciousness. Although we could argue that this is a special form of vitalism, it is commonly referred to as 'dualism.'

In the 20th century and beyond, the concepts of vitalism (and dualism) have been gradually undermined as advances in biology, biochemistry, physiology and other life sciences have provided more observable and objective explanations for life processes. This process has its roots in the 19th century in the ability to produce organic compounds (such as urea) in a laboratory, the discovery of enzymes and their mechanisms of action, and the discovery (and later understanding) of the role of DNA in reproduction and genetic functions. Darwin's theory of evolution provided another 'non-vital' explanation of the multiplicity of plant and animal species. Although aspects of all of these discoveries remain a work in progress, with much still to explain, processes previously attributable to 'vital' forces have systematically been replaced with knowledge based on chemistry and other objective ideas.

For example, it is clear that the scientific understanding of 'forces' has changed. From Empedocles' love and strife to the life forces of vitalism, a force was considered to be something non-physical and God-like – basically a belief system.

In contrast, today, forces are concepts of physical reality that exist in nature and can be observed and quantified. They are found in living and non-living things and are not considered to be the essence of life. Electromagnetic forces enable chemical bonds; osmotic forces control the flow of substances across membranes in living beings; gravitational forces act on all parts of living bodies; and so on. These forces are all part of life but are not exclusive to life nor defining of life.

In summary, there has been a distinct shift in the past several centuries in how life and life processes are perceived by people who now call themselves 'scientists.' As more of these processes have become defined and understood based on chemical and other objective physical phenomena, the default has become to attribute what is unknown not to mysterious forces but rather to science-based discoveries yet to come. Nonetheless there are still many people today, including some scientists, who believe that non-scientific or God-like processes are important in defining life.

CHAPTER 6

ERWIN SCHRÖDINGER

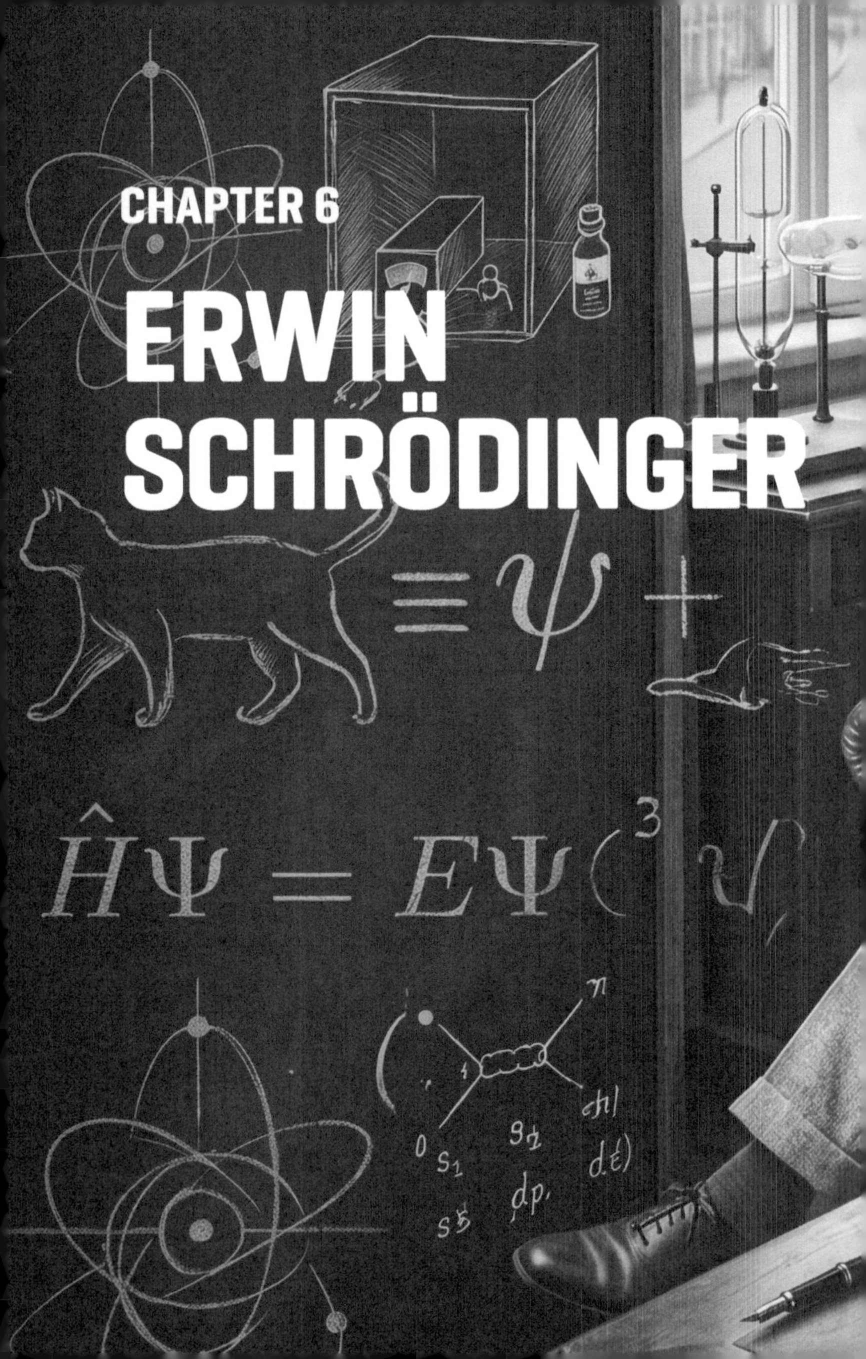

Quanten-
mechanik

In 1944, the renowned physicist Erwin Schrödinger wrote a short book titled *What Is Life?*[9] He was a Nobel Prize-winning physicist best known in his field for his wave equations relating to quantum mechanics. To the lay public he is probably best known for his whimsical thought experiment illustrating 'quantum superposition,' in which a theoretical cat contained in a closed box can be considered both alive and dead at the same time until someone opens the box and observes the cat ('Schrödinger's cat').* He was not a biologist and took pains in his book to apologize for his attempts to apply laws of physics to the controversial subject of the definition of life. However, no apology was needed: his book has become a classic and reverberates throughout this debate today.

Frustratingly, Schrödinger never quite writes an easily quotable definition of life in the book. However, his premise is that life differs from non-life in its ability to overcome (at least temporarily) the second law of thermodynamics, which states that in all natural systems, disorder (or entropy) increases with time. That is, the ability to increase or at least maintain order (rather than increase disorder or entropy) over long periods, and even across generations, is the essence of life.

---

* In quantum mechanics parlance, resolving a superposition through observation is called collapsing the Schrödinger wave equation.

According to Schrödinger's analysis, living things accomplish this remarkable physical feat using two special processes:

1. Ingesting 'negative entropy'

2. Creating 'order from order'

What he means by 'negative entropy' is that living organisms need to take in (by whatever means) substances that have negative entropy to offset the natural production of entropy associated with all natural processes, including living. That is, organisms, including humans, ingest foods (with low or negative entropy) and excrete waste and heat (high entropy) while maintaining a positive balance of order over lifetimes. He states that these foodstuffs must be "...extremely well-ordered state of matter in more or less complicated organic compounds..."* These compounds are stores of chemical energy, which are released through metabolism.

Thus, it is not simply metabolism but rather the metabolism of special negative-entropy substances that distinguishes the living from the non-living. Death occurs when these substances are no longer available or for whatever reason can no longer be ingested. Entropy and the second law of thermodynamics then prevail.

---

* Erwin Schrödinger, *What Is Life?* (Cambridge: Cambridge University Press, 1944), page 73.

A key aspect of Schrödinger's theory is the ability of organisms to create order from order rather than the more common physical phenomena physicists typically encounter of creating order from disorder. In 1944, biologists knew that the structures in the nuclei of living cells, called chromosomes, contained genetic code. They did not know the full chemical nature of genes, nor the exact mechanism of protein creation and DNA replication through a double-helix structure.

Nonetheless, through his review of biologists' research, Schrödinger knew there were a small and fixed number of chromosomes in each species that were contained in each cell, whether an organism was single-celled or multicellular.[†] He also knew that somehow the information necessary to create all aspects of any organism was contained in this small number of particles (chromosomes). He was amazed at how the small number of atoms that constituted a chromosome could generate such an enormous amount of information controlling all the structures and processes of any organism. This meant that a chromosome had to be a very orderly and complicated molecule itself: thus, the full order of a complete organism was created from a very orderly molecule. Hence, order from order.

---

† He mistakenly said that the number was 48 in humans rather than 46. In 1944, at the time of his publication, 48 was the number most commonly used. The correct number wasn't established until 1956.

This process is distinctly different from what is generally seen in the natural world, where order tends to be created from disorder. That is, material substances (such as all solids, liquids and gases) consist of an astronomical number of atoms that in turn create the astronomical number of molecules making up any substance. Individually these atoms and molecules act in a very random and disordered manner; however, in the aggregate, the laws of physics determine that they act in a predictable fashion. For example, the molecules of any gas move about randomly and chaotically (i.e. very disordered), yet in aggregate they follow precise laws of physics in predicting such variables as pressure and volume under various conditions. Hence, order from disorder. For a physicist, the idea of order coming from disorder was something commonly encountered. In contrast, there is nothing disordered in the way the chromosome defines all aspects of an organism. That struck Schrödinger as very unusual and amazing.

The ability of organisms to evolve in response to environmental conditions through natural selection is a component of most definitions of life. Schrödinger understood that mutations enable Darwinian evolution. Citing research on mutations in *Drosophila* flies, he noted the ability to identify the precise (and very small) location on the chromosome of each mutation. This was further evidence of his 'order from order' principle of life in that a very small number of atoms enable a major change in the 'program of life' for an organism that goes beyond the current generation.

Not only can these small groups of atoms propagate through future generations but they also provide the mechanism for entire species to evolve through natural selection.

Schrödinger equated mutations with quantum mechanics in that, in modifying a gene, they are akin to quantum leaps. There is no state of the gene in-between the mutation and the original form, just as there is no in-between state in energy levels as an electron moves from one level to another in an atom. Schrödinger predicted that the future elucidation of genes and DNA would explain the enormous amount of information contained in a single DNA molecule. He correctly described the material of chromosomes as aperiodic crystals, which have much more potential informational power than the periodic crystals found commonly in nature (see the Glossary for explanations of these terms).

Back in his home territory of physics, Schrödinger is best remembered for his wave equation. This was more than a successful mathematical description of elementary particle events. It was a giant leap in thinking and intuition and became the basis for all quantum mechanics. Similarly, his foray into biology was a masterstroke of brilliant proportions that brought this creative thinking to bear on events outside his core area of expertise. His definition of life as defying entropy remains a piece of the puzzle today. His descriptions of biological processes (including metabolism), genetic theory, chromosomal structure and evolution using physical laws were valid at a time when relatively little was known about all these subjects. His book was less than 100 pages long, yet it is perhaps the best example of the principle of 'order from order.'

But is this the end of the story? Does defying the second law of thermodynamics define life? No. There are many examples of non-living processes in nature that also defy the second law: for example, crystal formation in liquids or star formation in the cosmos. Although Schrödinger was decades ahead of his time in giving us a vital new perspective on this problem, it is not sufficient to define life solely on the basis of defying the second law of thermodynamics. However, defying the second law by using the information in a highly ordered 'aperiodic' crystal to define both the structure and processes of an entity could constitute a definition of life. Schrödinger never explicitly states that as a definition of life, nor any other definition. In fact, his entire book is more an attempt to explain life rather than define it.

CHAPTER 7

THE MODERN SEARCH FOR A DEFINITION

Debates between scientists, philosophers, engineers and many others continue today regarding the definition of life. Although people from every field and those with many different motivations are participating in this debate, three groups need such a definition to achieve their goals:

- Those searching for extraterrestrial life
- Those attempting to create artificial life in a laboratory
- Those interested in determining the origin of life

For these three groups, success will be difficult to achieve unless there is some consensus on a definition. For this and many other reasons, professional meetings and publications devoted to finding one continue to occur. Two very memorable such meetings occurred in the first decade of this century.

The first – called the Workshop on Life – was organized by the Catholic Church and held in Modena, Italy, in September 2000. Its proceedings were later published in a book called *Fundamentals of Life*. At the meeting, scholars from many fields of science and religion came together to present papers and debate this subject. The opening sentence of the conference proceedings basically defines the challenge not only of the conference but of my book: "Life is something very particular, a category which everybody understands, but which cannot be defined exactly at the present stage of human knowledge."[10]

Before delving into attempts at a definition, the book's editors pointed out three characteristics of life that are overarching:

- **Complexity** – even for the simplest single-celled organisms

- **Genetic code** – highly sophisticated elements of information or even knowledge

- **Purposefulness** – this is seen in life's defiance of the second law of thermodynamics, but there is also a need for new laws to account for higher levels of purposefulness

Following this introduction, the editors present 78 definitions of life from the various participants, who were mostly world-renowned scientists varying in their backgrounds, education, research experience and nationalities. Some of these definitions are long, and some are more like explanations of why there can't be a definition. Following are those definitions that can be summarized in a single sentence:

1. Life is the process of existence of open non-equilibrium complete systems that are composed of carbon-based polymers and are able to self-reproduce and evolve on [the] basis of template synthesis of their polymer components. (Anatoly D. Altstein, Institute of Gene Biology, Russian Academy of Sciences, Moscow, Russia)

2. Life is a system of self-organization, self-replication, evolution through mutation, metabolism and concentrative encapsulation.* (Gustaf Arrhenius, Scripps Institution of Oceanography, La Jolla, California, USA)

3. Life is a self-sustained molecular system that transforms energy and matter, thus realizing its capacity of replication with mutations and anastrophic evolution.† (Herrick Baltscheffsky, Department of Biochemistry, Arrhenius Laboratories, University of Stockholm, Stockholm, Sweden)

4. Life is a set of symbiotically linked molecular engines, permanently operating out of equilibrium, in an open flow of energy and matter, although they recycle a great deal of their own chemical components through cyclic chemistry. (Laurent Boiteau, Chemistry Department, University of Montpellier, Montpellier, France)

5. Life is a chemical system capable of transferring its molecular information independently (self-reproduction) and also capable of making some accidental errors to allow the system to evolve (evolution). (André Brack, astrobiologist, Centre National de la Recherche Scientifique, Orleans, France)

---

* This is referring to cell membranes.

† The term 'anastrophic evolution' is not widely used or accepted in science. It implies some kind of inversion or reversal of a trend, such as simplification rather than more complexity, or a reversal of environmental conditions.

6. Life is a non-equilibrium chemical system that contains organic polymers, that can reproduce itself, that can metabolize by itself and that is segregated from the environment. (Roger Buick, astrobiologist, School of Geosciences, University of Sydney, Sydney, Australia)

7. Life is what is common to all living beings.* (Christian de Duve, Institute of Cellular and Molecular Pathology, Brussels, Belgium)

8. Life is any homo- or heterotrophic cellular irreversible heat engine, or their assembly, that carries instructions for its function, reproduction, topical location, individuality and life cycle. (Frederick Roland Eirich, physical chemist, Hightstown, New Jersey, USA)

9. Life consists of metabolism, autonomy, a steady state, negative entropy, reproduction, development and learning. (Avshalom Elitzur, physicist and philosopher, Unit for Interdisciplinary Studies, Bar-Ilan University, Ramat-Gan, Israel)

10. "Life is what competent biologists consider as life" (quoting Theodosius Dobzhansky).† (Imre Friedman, Florida State University, Tallahassee, Florida, USA)

---

* After that one sentence, he goes on to try to convince his readers that it is not a tautology.

† Theodosius Dobzhansky was a renowned Russian American geneticist and evolutionary biologist.

11. Life is matter that makes choices, binds time and breaks gradients. (Ricardo Guerrero, Department of Microbiology, University of Barcelona, Barcelona, Spain, and Lynn Margulis, Department of Geosciences, University of Massachusetts, Amherst, Massachusetts, USA)

12. The functions that are called life are metabolism, growth, and reproduction with stability through generations. (Romeu Cardoso Guimarães, physician geneticist, Federal University of Minas Gerais, Belo Horizonte, Brazil)

13. Life is simply a state of organized instability. (Remy J. C. Hennet, geochemist and hydrogeologist, Bethesda, Maryland, USA)

14. Life is a system that has subjectivity. (Kunio Kawamura, Department of Applied Chemistry, Osaka Prefecture University, Osaka, Japan)

15. Life is metabolism and proliferation. (Lajos Keszthelyi, Institute of Biophysics, Hungarian Academy of Sciences, Szeged, Hungary)

16. Life is a population of functionally connected, local, non-linear, informationally controlled chemical systems that are able to self-reproduce, to adapt and to coevolve to higher levels of global functional complexity. (Günter von Kiedrowski, Chair of Organic Chemistry, Ruhr University, Bochum, Germany)

17. Life is a new quality, characterized by the ability of temporal self-maintenance and self-preservation, brought upon an organic chemical system by a dialectic change resulting from an increase in the quantity of complexity of the system. (Vera M. Kolb, Department of Chemistry, University of Wisconsin-Parkside, Kenosha, Wisconsin, USA)

18. Life is an organized form of intensified resistance to self-propagating processes of destruction. (Vladimir N. Kompanichenko, astrobiologist, Russian Academy of Sciences, Khabarovsk, Russia)

19. Life is any system that creates, maintains and/or modifies dissymmetry. (Wolfgang E. Krumbein, Institute of Chemistry and Biology of the Marine Environment, Ossietzky University, Oldenburg, Germany)

20. Life is the form of existence of the substance capable of self-reproduction and maintenance of permanent metabolism with the environment. (Igor S. Kulaev, biochemist, Russian Academy of Sciences, Pushchino, Russia)

21. A terrestrial living entity is an ensemble of molecular-informational feedback-loop systems consisting of a plurality of organic molecules of various kinds, coupled spatially and functionally by means of template-and-sequence directed networks of catalysed

reactions and utilizing, interactively, energy and inorganic and organic molecules from the environment. (Noam Lahav, chemist, Hebrew University of Jerusalem, Rehovot, Israel)

22. Life is an auto-regulated ordered system that can persist through time facing entropic equilibrium, either by remaining a single entity or by replication, and that can evolve. (Purificación López-García, evolutionary biologist and microbial ecologist, National Centre for Scientific Research, Orsay, France)

23. A living entity is an ensemble of molecules that exhibits spatial organization and feedback loops in using materials and energy from the environment for its growth and reproduction. (Shlomo Nir, biophysicist, Hebrew University of Jerusalem, Rehovot, Israel)

24. Life is a system that can reproduce itself by using a genetic mechanism. (Haruhiko Noda, chemist and biophysicist, University of Tokyo, Tokyo, Japan)

25. Life is the ability to reproduce and evolve. (Tobias Owen, Institute for Astronomy, University of Hawaii, Honolulu, Hawaii, USA)

26. Life on Earth is replication, variation and selection of individualized membrane-separated systems of nucleic acids that are served, structurally and catalytically,

> by nucleic-acid-encoded proteins and their complexes with other compounds. (Alexander S. Spirin, biochemist, Institute of Protein Research, Pushchino, Russia)

The above list is probably the best argument against there ever being a consensus definition. What was most interesting to me in reading these definitions was the backgrounds of the individuals who submitted them. These were renowned scientists from many different countries, in many different fields, highly published, almost all of them academically based. A large number of them were engaged in studies of the origin of life or the search for extraterrestrial life, both of which have need of such a definition. Also interesting is the fact that this conference was organized by the Catholic Church. Although the proceedings had a great deal of religious content and thoughtful input from clerics and others from religious backgrounds, the brief definitions of life listed above are highly science-based and generally lack the vitalism and spiritual threads of religious views of life.

As to the definition of life itself, it did not appear to me that there was any attempt to achieve any kind of consensus. Certainly, if the list above characterizes the entire meeting, consensus would have been impossible to achieve. To me, the most resonant comment came from Herbert Shaw, a brilliant geoscientist at the US Geological Survey: "Life is like consciousness: If you think you can explain what it is, you've got it all wrong."[11]

A second memorable meeting on the definition of life occurred in Paris later in that decade, in 2008. With the title Defining Life, it brought together many of the premier scholars on this subject.[12]

There were two general themes that permeated this conference. The first was that if the goal was to achieve a consensus, then the definition would need to be as simple as possible. The second theme was that it would be impossible to achieve such a consensus.

One possible 'simple' definition was discussed that came from the NASA Exobiology Program in 1992: "Life is a self-sustained chemical system capable of undergoing Darwinian evolution."[13] NASA's interest in this topic stems from its involvement in the search both for extraterrestrial life and for clues to the origin of life.[14] This definition is also widely quoted in the literature.

Although this one-sentence definition is simple, it is not useful for many purposes. For example, how would we use it to confirm life on Jupiter's moon Europa, which is considered a possibility for extraterrestrial life. Europa has water and some conditions similar to places on Earth where certain organisms thrive. However, even if scientists detected chemical reactions characteristic of earthly organisms, how would they confirm how sustained those reactions are? In other words, how long is 'sustained'? Even more problematic, how would they confirm Darwinian evolution?

Or even the 'capability' for it? Before anyone would believe that life had been discovered on Europa, a lot more evidence would be required.

So, although the NASA definition seems to be a true statement about life, it doesn't seem sufficient to define it. We would need to better define the terms it uses to allow confirmation that they had been met in any instance. For example, regarding the Darwinian evolution 'capability,' would it be necessary to say that the entity contains a genetic material that undergoes a sufficient number (but not too large a number) of naturally occurring modifications (e.g. mutations) that would provide enough variation regarding reproduction capability to subject the entity to natural selection from its environment? Or would we somehow need to observe the evolution directly? In either case, this simple definition of life has suddenly become much less straightforward and much more open to debate.

Most of the papers from this conference that attempt definitions resort to creating minimal lists of features that define life. These features include environmental isolation through encapsulation (i.e. an entity is composed of cells with membranes), self-reproduction, ability to evolve by interacting with the environment, self-maintenance through metabolism, information-capable genetic material, ability to exploit external energy, autopoiesis (the ability to self-regenerate through a network of processes) and response to stimuli. The problem is that no two of the definitions share the exact same list of features (in the next chapter, I will discuss one such list that I think is as good as any from this conference and explore its problems as a definition).

Regarding the second theme of this conference – that it will be impossible to ever obtain a consensus on a definition – the above discussion reflects the difficulty. There may be so many variables to define, each of which may have multiple options, that achieving a consensus is a remote possibility. But, besides this complexity, there are some other reasons (discussed at the conference) that a consensus may be impossible.

One is that there are many different types of definition! One conference paper listed the following types of definition:

- The synonymous method
- The method of analysis by genus and differentia (specific features in the genus)
- The relational or synthetic method
- The implicit method
- The (so-called) regular method[15]

Although the paper does not provide examples of these types as they pertain to the definition of life, it is inherently clear that different authors have used different types of definitions that are not necessarily incompatible with each other. Also, there are subtle differences between defining 'life,' 'living' and 'living organisms.' Concepts of 'life' can be more general, philosophical and abstract, whereas 'living' implies a process. Attempts to define 'living organisms' tend to invite lists of features. Thus, there may reasonably be more than one definition, with each one serving a different purpose. The problem, potentially, is getting a consensus not just on one definition but on multiple.

Finally, and most problematically, a couple of the papers suggest that the difference between life and non-life may not be binary. That is, there may be a continuum between non-life and life that has no clear separation between the two. If we consider the circumstances around the origin of life, there clearly was a pre-life period in which there existed various precursors to the molecules required for life. Somehow, these molecules and the chemical reactions in which they were involved changed in a manner that led to life. Those changes happened over time and were undoubtedly incremental. The question is whether this happened over a continuum of changes, where it was unclear exactly when the crossover to life occurred, or whether the threshold was crossed abruptly. Later in this chapter we will encounter some potential intermediaries between living and non-living, but there is no evidence that these were present at life's beginning.

There is much speculation that this borderline state, if it existed, was somehow related to RNA. We know that RNA is self-replicating, although not in the same way as DNA. We also know that RNA can act as an enzyme in many chemical processes, including protein production. We know there are RNA viruses today, but they can only live by taking over the reproductive machinery of a DNA cell. We know that RNA can evolve through natural selection. So, was there an early RNA-based entity that was somehow intermediate between living and non-living in some way? We don't have any evidence of such an entity. Perhaps if there was, it was sustained for only short periods of time and was ephemeral in its existence. Or did it constantly change its key characteristics back and forth between something alive and something not considered alive? Or was it something that simply defied definition with respect to life?

Thomas Cech received the Nobel Prize in Chemistry for his studies of RNA. In his book *The Catalyst*,[16] he devotes an entire chapter to how a self-sustaining RNA-based organism could have emerged from the chemicals and conditions present on Earth 3.8 million years ago, when life originated on Earth. It would have been enclosed in a lipid membrane, which many definitions of life today would require. In Cech's description, such an organism would not necessarily be considered intermediate between living and non-living unless the definition required DNA, which all known living organisms today require. Even if we consider such an RNA-based organism as living, there were likely intermediate states in the run-up to it. We would still have the same problem of differentiating this RNA-based life from non-life, but at an earlier step in the process.

That brings us to the question of viruses. Are they an example of something intermediate between living and non-living? To answer this question, it is first important to clear up a basic misunderstanding that often permeates this debate: what is the definition of a virus?

The life cycle of a virus consists of two phases:

1. When it exists outside a living cell

2. When it exists inside a living cell

During the phase when it exists outside a living cell, it is called a virion. A virion consists of the genetic material of a virus, either DNA or RNA, surrounded by a protective protein shell called a capsid. Some capsids are further enclosed in a lipid membrane. The entire purpose of the virion is to enter a living cell (always DNA-based) of a host organism. Typically, as part of the capsid and/or lipid membrane are protrusions or spikes that engage a living cell upon contact and facilitate penetration of that cell's membrane, so the virion can enter the cell. Virions can survive in many environments outside cells, including blood, water particles and other surfaces. However, they cannot replicate in this phase, nor do they metabolize any substances. They would not fit anyone's definition of being alive in themselves. Could a virion be considered an intermediate form of life? That's a philosophical question and can be answered any way you like.

Once a virion enters a cell, its genetic material activates a number of chemical processes that, in essence, take over the reproductive machinery of the living host cell to produce more virions using structures, energy and substances provided by the host cell. In some cases, the virus's genetic material also becomes incorporated into the DNA of the host cell. Technically, the definition of a virus includes both phases of its existence.

During the COVID pandemic, we saw many beautiful domed and spiked images of the virion form of the virus – not pictures of the virus itself. It would be analogous to seeing a picture of a larval or caterpillar phase of a butterfly or moth. There really is no way to photograph a virus in its host phase.

There are many types of viruses. Each type typically attacks primarily a single or small number of species of plant, animal, fungus, bacterium or archaeon. Many viruses mutate in a manner that allows crossover to other species. This is an example of virus evolution. In rare cases, they can even infect another virus while it is inside its host cell.

The question of whether a virus is alive or an intermediate form of life must consider the entire life cycle of that virus.

The answer, then, depends on the definition we choose. Under most definitions, the answer is that a virus is alive and is not an intermediate form. However, if we choose to define life in terms of organisms requiring their own lipid membrane (rather than a borrowed one) or cell, or their own machinery to create proteins, or their own energy sources, then they would not meet that definition. They would be considered non-living or perhaps intermediate. This takes us back to the second theme of the 2008 Paris conference: there will never be a consensus on a definition of life.

There have been no major international conferences devoted to defining life since the two referenced here. However, it remains a topic or theme of discussion in meetings devoted to astrobiology, the origins of life and/or artificial life. The status of the definition remains the same: no consensus.

In 2025, a more modern approach to finding a consensus definition of life was published titled "What Lives? A Meta-analysis of Diverse Opinions on the Definition of Life."[17] The authors applied artificial intelligence to the problem. Specifically, they used three popular large language models (Claude, GPT and Llama) to semantically compare definitions provided by 68 invited scientists and experts in various scientific disciplines from multiple institutions and countries. Each was requested to submit a definition no longer than three sentences. None of the submitters saw any of the other submissions until after they had submitted theirs.

Using the three AI tools plus somewhat complex mathematical analysis, the authors then quantified the correlations of the AI-based semantic couplings of these 68 definitions to determine whether and how well they clustered into common themes. Certainly, they clustered, but not into a single representative definition. Rather, they fell into an overlapping continuum of eight definitions representing a range of scientific thought. In other words, there was no single definition but rather multiple but conceptually related definitions that could serve to bridge rather than fragment the scientific community's thinking on this topic.

The authors point out that the purpose of this paper was not necessarily to come up with the definitions themselves but to demonstrate what they considered to be a new and valid methodology: using AI to resolve differences of opinion between scientists on any question (another example is the definition of cognitive processes). They point out that the number of scientists chosen (68) was not necessarily sufficient to be representative of the breadth of expert opinion existing in the world. In other words, a single definition (or even group of definitions) may be even more elusive than the paper's findings suggest.

The 68 submitted definitions do not vary dramatically from the 78 published following the 2000 Modena conference (and there is no overlap between the participants in the two groups). Virtually all the same possible components of a definition are included in one or more of the eight

AI-determined definitions. The 2025 paper's spectrum of definitions reinforces three notions: it is impossible to achieve a consensus on a single definition, multiple definitions are needed, and there may not be a distinct and definable separation between life and non-life but rather a non-binary continuum.

CHAPTER 8

THE SEVEN PILLARS OF LIFE

COMPARTMENTALIZATION
ENERGY SOURCE
REGENERATION
ADAPTABILITY
SECLUSION

Daniel E. Koshland Jr. was a pioneering biochemist who served ten years as the editor of *Science* magazine. In 2002, he published an article in *Science* titled "The Seven Pillars of Life."[18] This was his attempt to achieve what he acknowledged all others had so far failed to do: define life. He eschewed a simple definition because his experience was that all such previous attempts had only led to disagreements and exceptions. Instead, he attempted to be comprehensive. Even so, he admitted that his definition only applied to life as we know it on Earth. He also reserved the right to alter his definition, if and when a non-conforming example was discovered elsewhere in the universe.

Here are his seven pillars:

1. **Program** – For life on Earth, this means DNA. DNA defines both the ingredients needed for sustained life and the kinetics between those ingredients.

2. **Improvisation** – This is basically evolution through natural selection based on mutations in DNA so that organisms can adapt to environmental changes and threats. (Others use the term 'adaptability' for this but Koshland uses that term differently, as we'll see.)

3. **Compartmentalization** – Organisms have cells with membranes so that they can retain the appropriate concentrations of chemicals to allow appropriate kinetics.

4. **Energy source** – This is to sustain metabolism and resist entropy. Ultimately the energy source is the sun, but transformed into the energy-producing components of each cell, such as those contained in mitochondria (considered further in the next chapter).

5. **Regeneration** – This refers to the constant repair and replacement of chemicals, cellular structures and tissues. It is imperfect in a single organism, leading to ageing and death. However, through reproduction, it starts anew with the next generation. (Here, Koshland seems to combine what others call 'sustainability' and 'reproduction' into a single feature.)

6. **Adaptability** – This is different from Pillar 2, Improvisation, in that it refers to individual organisms' short-term responses. It includes responses to stimuli through pain, thirst, hunger and other short-term reactions to the environment.

7. **Seclusion** – This is the ability that enables hundreds or even thousands of chemical reactions in a single cell to occur without each one interfering with the others through random collisions with the various chemicals. It happens primarily through the ability of catalysts to select the exact molecules needed for a reaction out of the 'chemical soup' environment of a cell. (This inclusion in a definition of life is unique and reflects the biochemist speaking.)

Koshland wrote his definition almost 25 years before this writing. I single it out simply as an example of the most common type of definition we see in the literature – a list of multiple features that define life. This is a heavily cited article by a well-respected biochemist, but we could hardly say that it is a widely accepted definition today. It's just one of many, including those in the 2008 international Paris conference discussed in the previous chapter.

Certainly, the seven pillars are all generally accepted features of living organisms, although we could (and many have) quibble about the exact meanings of each of them. But there are many other features of living organisms not mentioned, such as the ability to eliminate waste, the influence of epigenetic factors in gene expression, the use of ribosomes for protein creation, the use of adenosine triphosphate (ATP) for energy, the role of RNA, the negative entropy of food sources mentioned by Schrödinger (see *Chapter 6*) and many others. There is no broad model or conceptual basis for choosing those particular seven. A later article by Y. N. Zhuravlev and V. A. Avetisov criticizes Koshland, stating, "it seems unlikely that properties of 'compartmentalization, program and seclusion' are the features of non-differentiated primary life."[19] I'm not sure I agree with that critique but I mention it to illustrate the difficulty of ever reaching any consensus.

The other problem is how we can apply this definition, or any definition that is a list of features, to specific cases. It is still unclear how the definition relates to viruses, for example (see the previous chapter). Viruses would only meet all seven pillars if it were accepted that some of the seven are provided by the host cell and not the virus itself. I don't know how Koshland would have responded. An RNA-based origination of life would also not meet the Koshland definition.

CHAPTER 9

WHAT IS LIVING?

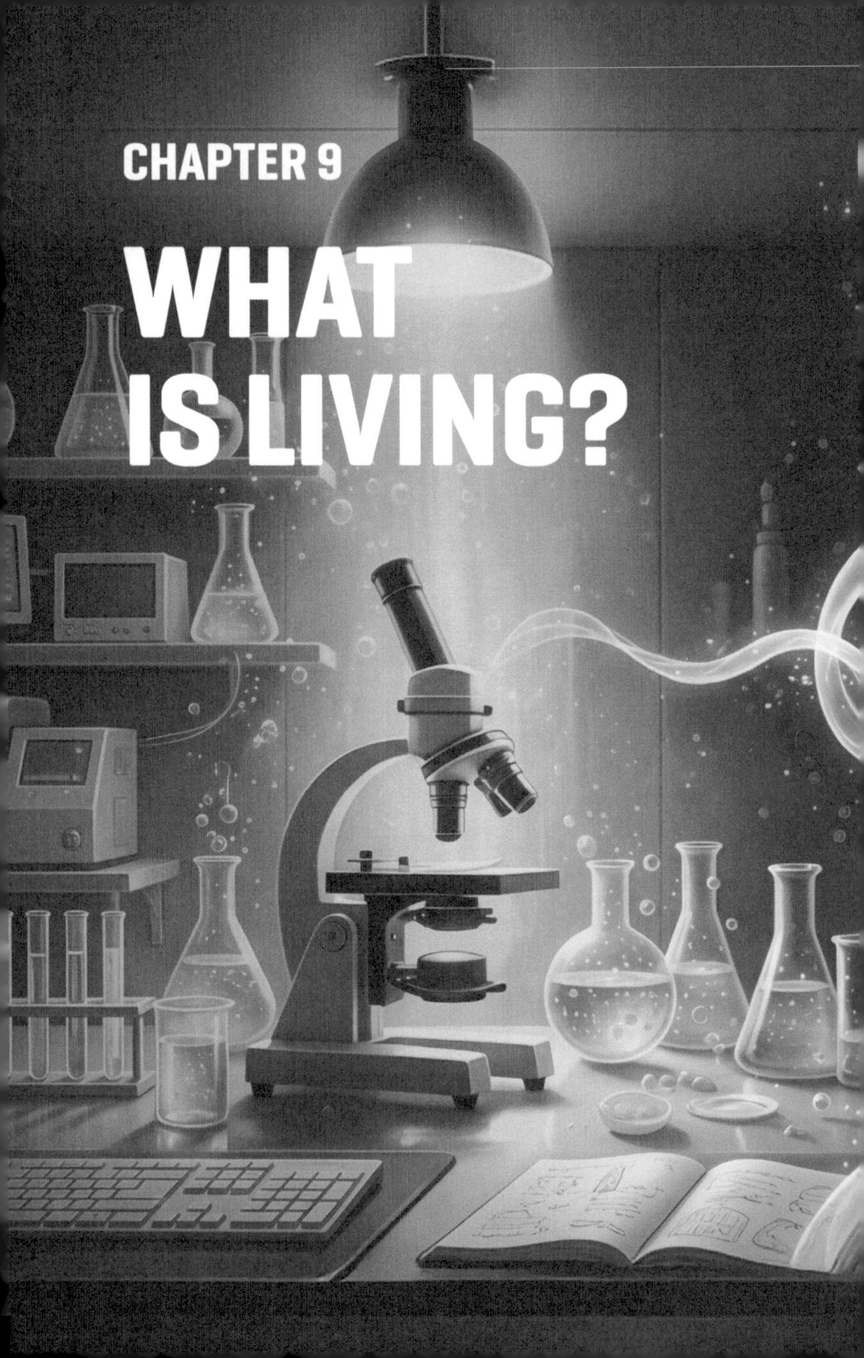

In *Chapter 7* we saw that viruses are difficult to fit into any definition of life. But there are other examples that are also difficult to classify.

As we saw, a virus in its virion stage would not be considered alive, since it can't reproduce or metabolize. But maybe that is irrelevant. Single phases in the life cycles of any organism should probably not be classified. That would include, for example, the caterpillar phase of the butterfly. I think if you asked anyone, including scientists, whether caterpillars are alive, there would be an overwhelmingly positive response, even though technically caterpillars cannot in themselves reproduce, evolve or do some of the other things those same scientists would include in their definition of life. Caterpillars certainly seem 'more alive' than virions in that they have their own cells with membranes, they metabolize, they respond to stimuli, and they perform cellular reproduction and other life functions. However, the caterpillar does raise some semantic issues for a definition that would withstand quibbling.

A similar concern comes from requiring the ability to reproduce as part of a definition. For most bisexual animals, it takes a male and a female to reproduce. That is, the individuals cannot reproduce by themselves although they are certainly alive. Thus, the requirement of reproduction applies to the species, not to the individuals. But mules (which technically are not a species) cannot reproduce either whether male, female or jointly. Certainly, mules are alive.

Without reproduction, there would be no life unless life emerges spontaneously repeatedly. There is no evidence that life has emerged more than once on Earth (see *Chapter 3*) and it definitely does not do so repeatedly. So, how do we include reproduction in a definition while also acknowledging a mule and a caterpillar as living beings? It would require semantic gymnastics, the kind that gets in the way of creating definitions that are widely acceptable. Or, alternatively, can we define life without requiring reproduction? That would seem bizarre to me. Note that in the previous chapter, Koshland did not give reproduction its own pillar, but combined it with 'regeneration.'

What about prions? Prions are infectious agents that can spread from one organism to another, reproduce and evolve. But that's all that they can do. They are proteins that are folded in ways that induce other normal proteins, particularly in brains, to misfold, leading to disease. They certainly are derived from DNA, probably through some type of mutation, but they themselves do not contain DNA, RNA or anything considered an information template. They contain no membrane and do not metabolize. I don't think any definition would consider them to be alive, although they do have many features of living things. Are they in between life and non-life, early precursors of life, or simply prions?

A virus is a type of parasite in that, as we've seen, it depends on a host for its survival. There are many types of parasite, however, whose life cycles are quite different from those of a virus. Some of those parasites can live, reproduce and do everything required by most definitions of life. However, 'obligate parasites,' such as the organisms that cause malaria, cannot reproduce without borrowing some material or process from a host organism. In the case of malaria, that host is a mosquito. But malaria organisms do not have a similar life cycle phase to a virus where they are unable to metabolize. Again, every reasonable definition of life would consider obligate parasites to be alive as long as that definition did not preclude entities that required material or processes from hosts.

Speaking of parasites, there is an interesting organelle in all plant and animal cells called the mitochondrium. Mitochondria are essential for life in these organisms and certainly are not parasites. They provide the energy for cellular processes. However, they have their own DNA, which is different from the DNA of the cells they occupy. In other words, they have their own genes. They exist within a cell in vesicles containing their own membranes and perform all the other processes normally considered to define life, including reproduction, mutation and evolution.

Bacteria and archaea do not contain mitochondria. It is believed that in the evolution of eukaryotic cells (those cell types that contain nuclei and that are the basic building blocks of plants, animals and other more complex organisms), either a bacterial species or an archaeal species ingested another bacterial species and incorporated it for future generations as mitochondria. It is speculated that this initial combination was a parasitic relationship that evolved into a symbiotic one. So, in a sense, mitochondria could themselves fit the definition of a living thing. A similar story would apply to the chloroplast organelles in plants, which produce oxygen. They were also probably ingested bacteria originally.

This brings us to another topic about cellular life that raises some complicated questions. For decades, we have been able to grow cells from various parts of organisms, including humans, in the laboratory. These cells can be of many different types, including tumor cells. Typically, cells are grown as 'cultures' (two-dimensional arrays on a growth medium of some kind) and they are used for many types of research and medical practice, such as to observe their reaction to various drugs. For example, doctors might culture tumor cells in a lab from patients' biopsies and then test the cells' sensitivity to various chemotherapy drugs to work out the best drugs to prescribe. These cells are certainly 'alive' in the sense that they have their own membranes, they can grow additional cells in the culture, they metabolize, they contain DNA and they perform most, if not all, of the other functions contained in most definitions of life. They could even evolve through mutations and environmental selection (e.g. from alterations in their culture medium). In themselves, they are not complete organisms, or even tissues, and they are kept alive artificially by humans. This lack of viability outside a lab may preclude many from considering them alive. On the other hand, it depends on your definition.

The most famous cell cultures are called HeLa cells. These cells originally came from a cervical tumor biopsy of a woman at the Johns Hopkins Hospital in 1951. This was during the early period of cell culture technology research and scientists were having difficulty sustaining cell cultures in the laboratory. However, amazingly, this particular cell culture grew indefinitely,* allowing multiple batches from the original culture to be sent to scientists all over the world for use in numerous and important research developments. Descendents of those original cells are still in use worldwide today. In essence, they are immortal!

They are called HeLa cells because the procedure at Johns Hopkins in those days was to label the cultures based on the name of the individual from whom they were taken – in this case, Henrietta Lacks. Years later, this inspired a well-researched book, called *The Immortal Life of Henrietta Lacks*, that told the incredible tale surrounding HeLa cells.[20] Henrietta Lacks was a Black woman living near Baltimore, Maryland, in the USA, who was diagnosed and treated for cervical cancer at the Johns Hopkins Hospital.

---

* We now understand the genetics of human cells, including tumor cells. Typically, tumor cells contain many more mutations than a person's normal cells. Some tumor cells, such as those of HeLa, contain mutations that prevent the normal ageing and death of cell lines as they reproduce. Most cell cultures do not live indefinitely.

She subsequently died of her illness. She and her family were totally unaware until many years later that her cells had become famous in scientific circles and were in use throughout the world. The book describes the controversies regarding the treatment of Black patients at that time at Johns Hopkins; issues regarding informed consent; the legal, financial and ethical issues surrounding this case; and the history of HeLa cells. Can we regard Henrietta Lacks as immortal? I don't think so. Are her HeLa cells immortal? Yes, at least so far after 75 years. Does that mean they are alive? I think so.

In recent years, we clever *Homo sapiens* have taken this cell culture ability much further. We can now create in the lab miniature organs, called organoids, that are much more than a simple array of one type of cell. They are three-dimensional tissues containing multiple types of cell of a single organ type functioning together very similarly to natural organs. We now have organoids that simulate, in a lab, parts of a brain, intestine, kidney, lung and other organs. These organoids are quite small, usually no more than a few cubic millimeters, and do not contain all the cell types in a natural organ. However, they perform enough of the functions of their organ type to provide valuable research. For example, one brain organoid, called DishBrain, has been taught to play a version of the video game *Pong* by being connected to a computer chip.[21]

Organoids are usually grown from pluripotential stem cells, with the process directed using techniques similar to those naturally occurring in the embryonic growth of organs. Organoids certainly fit the definition of life better than cell cultures, but for those who question whether cell cultures are alive, the fact that organoids contain multiple types of cells may not be enough to convince them that the threshold for being alive has been passed. That question will remain even if we finally produce complete organs in a lab. They still won't be viable organisms. And they still will need us humans to create and maintain the environment in which they live. Would your definition then rule them out as living?

Even more mind-blowing is that biologists are extending their techniques to create more complicated synthetic entities. The nomenclature is beginning to become blurry. These are not exactly organisms: they are still based on manipulating the pluripotential stem cells or embryonic cells of existing organisms. However, the goal is not to produce just a tissue or organ for research purposes, but to produce new types of viable entities that do not correspond to any existing species but can perform unusual and sometimes useful functions.

One article refers to them as synthetic living machines.[22] The authors describe how they extract cells from frog embryos and manipulate them in synthetic cultures. The cells develop into 'living' entities, each made up of thousands of cells, that the authors call 'xenobots.' These can swim on their own through various fluids using cells to propel them in ways that are not used in the frog from which they are derived. Specifically, the xenobots' cells have cilia that are normally used in the frog to move mucus but that are now used to swim. The cells have 'self-organized' in new ways in this synthetic process of development, and the entity does not conform to any naturally existing species.

The xenobots can 'live' for up to ten days by metabolizing food already contained in the embryonic cells from which they are derived, and longer if they are provided with an enriched culture media. They 'die' by shedding cells and disintegrating. They can repair themselves if lacerated while 'alive.' They can sense objects and move around them. They can aggregate with other xenobots and collectively remove debris from an area. Are they living? It sounds like it, but they don't reproduce.

Cell cultures, organoids and xenobots are created using cells taken from a living organism. They do not meet the definition of life created from scratch in a laboratory, which is a completely different endeavor. Those attempting to create life in a lab are seeking the answer to the question of the origin of life. Typically, these efforts attempt to create conditions (such as temperature) and environments and chemical mixtures thought to be similar to those on Earth when life began. Through various manipulations, such as electrical stimulation (mimicking lightning), and by adding catalysts that may have also been present, scientists observe how various chemicals (such as amino acids) and even nucleotides (the basis for DNA and RNA) and lipid membranes develop in these lab experiments. Although there have been some promising and suggestive results from these experiments,[23] so far sustainable life has not been created in this way.* If and when that occurs, there will surely be debates about whether what is created conforms to an accepted definition of life.

There is another way to create life from scratch in a lab other than mixing a lot of chemicals and exposing them to different conditions. It's called synthetic biology. Some may call this cheating. I call it mind-blowing.

We now have the ability to create DNA molecules, specifying the nucleotide sequence of the DNA almost as easily as typing out letters on a keyboard. Generally, this is used to create relatively short DNA sequences of genes, such as for use in gene editing. However, in 2010, a group of researchers from the J. Craig Venter Institute published a paper in *Science* magazine in which they synthesized the entire genome of a bacterium from a solution containing the four nucleotides of DNA.[24] This was done under the control of a

* It should not be surprising that, so far, no scientist has been able to create life in the laboratory using chemicals and conditions thought to be present 3.8 billion years ago. If we think of all the improbable things that would need to come together at the same time and place to create life, statistics alone would suggest that it might take thousands or even millions of years to occur even if all the ingredients were present. That works OK for geological time but doesn't fit very well into the professional motivations of any scientist.

computer using a digitized sequence of that organism's genome. The researchers then took that synthetic genome and transplanted it into cells with their natural DNA removed. This synthetically produced bacterial organism thrived and reproduced like its natural counterpart. That is, it lived. No one questions whether this synthetically created organism meets any of the many definitions of life.

But synthetic biologists do a lot more than this. The same Venter group, in association with others, also experimented with creating the shortest (i.e. fewest genes) known version of a living organism by systematically genetically engineering a bacterium genome to incrementally eliminate genes that it determined were not necessary for its sustained life. It ended with a viable organism of less than 500 genes compared to the more typical number of around 4,000 in most bacteria.[25] (Human cells, by contrast, have about 25,000 genes in their DNA.) Again, there is no question that this reduced-genome organism, although not natural, was alive.

Synthetic biologists also rearrange genes, subtract genes, add genes from other organisms, and create viable organisms to carry out unusual (often useful) functions that are not found naturally in nature, such as ingesting oil and waste products. Most of this is done by genetically engineering, copying and pasting, and adding and deleting existing natural DNA sequences.

Even more amazing is that synthetic biologists can now incorporate amino acids into proteins not normally used by existing natural organisms. Proteins are the mainstays of life, and all living organisms use DNA as a template to create protein-producing genes. The proteins these genes produce use just 20 different amino acids in various combinations to do everything needed for life. However, there exist in nature over 500 known amino acids. In rare cases, a few organisms use an additional one or two more naturally occurring amino acids, but almost all known life consists of just the same 20.* Synthetic biologists can now 'teach' DNA to incorporate over 200 more amino acids into proteins produced by genes. This has enabled the production of proteins with unique qualities (such as fluoresence or the ability to respond to specific wavelengths of light) and the creation of drugs with new properties.

Even more incredible is the ability to incorporate new nucleotides into a DNA or RNA sequence (each normally uses just four naturally occurring nucleotides). This means that, technically, the genetic material is no longer DNA or RNA. Yet it functions in a similar way in the synthetic organism.

Most would agree that all these synthetic organisms meet a definition of life – although it may be necessary to pick and choose a suitable definition (e.g. some require DNA and

* The other 480 known amino acids exists as free amino acids in water and on rocks, or in living organisms but not as a part of proteins.

some require other entities to occur naturally). It is possible to imagine synthesizing something that alters other fundamental functions considered necessary for life. Could there be a membrane-less type of cell? Or one that doesn't mutate and evolve? Who knows?

So far we're not doing a great job of defining life, but we're doing an excellent job of changing what it is that we're trying to define.

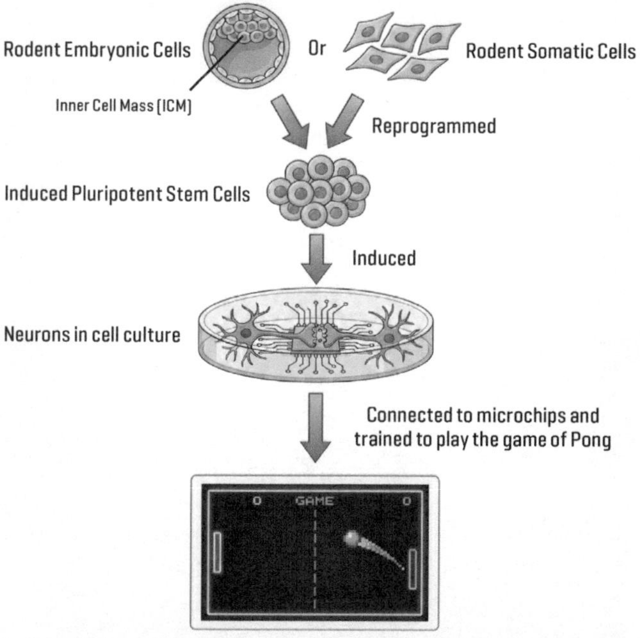

Synthetic biologists have taken rodent cells, reprogrammed them into stem cells, then induced them into brain cells and trained the brain cells to play Pong.

CHAPTER 10

CONCLUSION

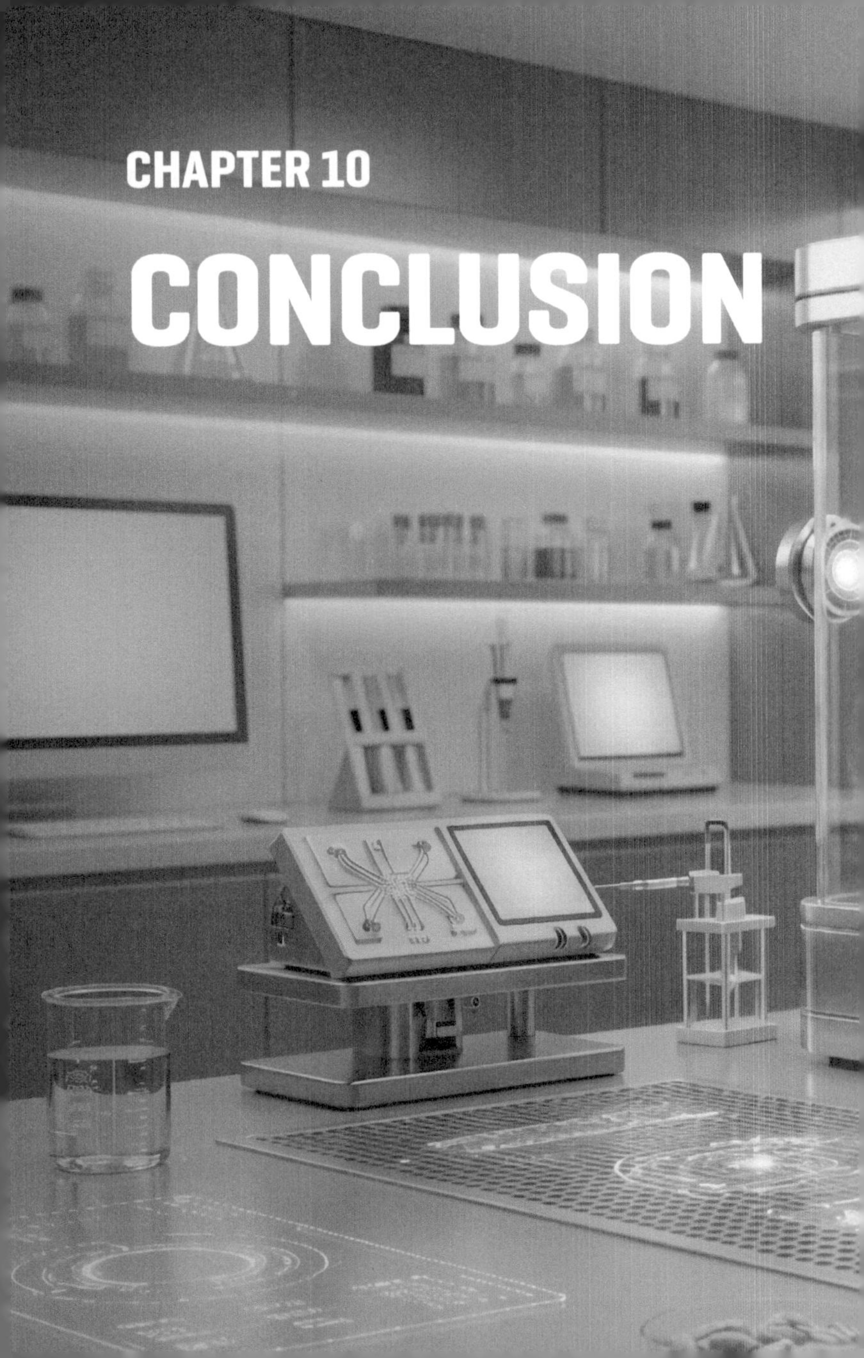

My conclusion is that the reason we don't have a widely accepted single definition of life is that there can never be one. The nature of the topic is simply too complex. There are too many variables to consider and too many types of definitions. The range of possible life forms defies a single definition, and these life forms are rapidly changing. There may be no clear distinction between life and nonlife, and there are too many types of scientists considering the problem. And that's all before we get the nonscientists involved.

So, when someone asks me "What is life?" I have a one-word answer:

AMAZING

EPILOGUE

My conclusion creates a bit of a problem for people whose life's work may depend on a widely accepted definition of life. How will I convince people that I've achieved my life's goal if I'm an astrophysicist looking for extraterrestrial life, or a biologist attempting to create life in a laboratory, or a scientist seeking the origin of life?

This problem is simply inherent in being a scientist. Very few things in science are certain and unchanging. Even if we had a widely accepted definition, it would change over time, just as science has moved on from its previously wide acceptance of vitalism a century or two ago (see *Chapter 5*). At any given time, any discovery will be greeted with "show me the evidence." That evidence will be judged by a panel of peers represented by the scientific community in publications, conferences and social media. Debate will surely ensue. A consensus (which is rarely 100% in science) may or may not follow regarding any particular claim.

It's what makes science interesting.

Astrobiologists, biologists and other scientists will live with it.

That's life.

GLOSSARY

This glossary defines terms as they are used in this book.

AMINO ACIDS

Small organic molecules that are the basic building blocks of proteins. Over 500 different types of amino acid are found in nature, but only the same 20 of them are used to create proteins in most organisms.

APERIODIC CRYSTAL

A chemical structure whose anatomical pattern does not repeat itself along the length of the structure, which may be quite large. Both DNA and RNA are considered aperiodic crystals. This accounts for their ability to serve as a totally flexible source of information for the genetic code. Periodic crystals, such as sugar and salt, could not serve this function.

ARCHAEA

One of the three major domains of living organisms in addition to Bacteria and Eukarya. They are single-celled like Bacteria but differ in the form of their cell membranes and chemical reactions. They are most numerous in extreme environments.

ASTROBIOLOGY

The scientific study of life in the universe. This multidisciplinary field focuses on the origin of life on Earth and where else it might exist.

ATP

Adenosine triphosphate. This molecule is the energy source that powers all chemical reactions and functions in all living organisms.

CATALYST

A chemical that speeds up a chemical reaction without itself being consumed in the reaction. Most biological reactions have catalysts. This evolutionary feature is vital to the efficient functioning of all organisms.

CHROMOSOMES

Packages of DNA containing the genetic material of an organism. The number of chromosomes in any species is constant and varies from one in most bacteria to hundreds in some plants. Humans have 46. The number of chromosomes does not correlate well with complexity.

CILIA

Hair-like protrusions from the surface of a eukaryote cell that collectively are used to move fluids [like mucus in the lungs] or the entire organism [like some single-celled eukaryotes] through a fluid.

COVID

An infectious disease caused by the virus called SARS-CoV-2 that caused the pandemic beginning in 2019.

DARWINIAN

Anything describing the theory of evolution by natural selection proposed by Charles Darwin in his 1859 book *On the Origin of Species*. An essential component of life is the ability of species to evolve over time because of mutations and other variations in their genetic code that better enable them to procreate in their environments.

DENISOVANS

A human species [*Homo denisova*] known only by a small number of fossils found originally in the Denisova Cave in Siberia. They interbred with both Neanderthals and *Homo sapiens* and went extinct about 30,000 years ago. Denisovan DNA is a small part of some of today's human populations. Whether they qualify as a separate human species is still being debated.

DIALECTIC

A method of reasoning through contradiction. For example, when describing evolutionary changes in organisms, a small change, such as a single DNA mutation, can lead to a huge change in the organisms. This would be called a dialectic change.

DNA

Deoxyribonucleic acid. It is the aperiodic crystal molecule that stores and transmits the genetic code [instructions] in all known living organisms. It consists of four nucleotides [adenine, thymine, cytosine and guanine] whose coding sequence determines the protein-producing genes necessary for life. Physically it exists as a double helix with adenine paired with thymine and cytosine paired with guanine.

ENTROPIC

Anything related to entropy – the tendency of all natural systems to move towards disorder.

EPIGENETIC

DNA or RNA factors that alter the expression or action of genes without modifying the gene sequence. For example, every cell in the human body has the same genes, yet, for example, the cells in the brain act differently from the cells in the heart. This is because of epigenetic factors that determine which genes are expressed [active] in each part of the body.

EUKARYA

The domain that consists of all eukaryote species.

EUKARYOTE

An organism consisting of cells that contain a membrane-enclosed nucleus. Complex organisms including plants, animals and fungi are eukaryotes, whereas bacteria and archaea are prokaryotes, lacking a nucleus.

GENE

A segment of DNA that specifies the coding of proteins or RNA molecules. Genes are the fundamental basis of all inherited traits.

GENOME

The complete DNA sequence of an organism including all genes and other coding and non-coding components, along with those elements thought to have no function ('junk').

HOMO ERECTUS

One of the earliest and longest-living human species. It migrated widely out of Africa to Asia and Europe and lived from 1.9 million years ago until about 110,000 years ago, when it became extinct. It is a part of the lineage leading to us, *Homo sapiens*.

KINETICS

The study of motion and its causes, effects of forces on moving bodies, and changes in physical and chemical systems.

LARGE LANGUAGE MODEL

Usually referred to as an LLM; a type of artificial intelligence system that is 'trained' on vast amounts of text, mostly from the internet, in order to determine statistical probabilities of the sequence of words. The result is a remarkable ability to produce human-like speech, generate computer programs, respond to questions, give advice, perform functions, engage in chat conversations and otherwise emulate a person.

LIPID MEMBRANE

An essential feature of all living cells that consists of a layer encapsulating all cellular functions. It is made up of special types of fats. In eukaryotes it also encapsulates the nucleus.

METABOLISM

The chemical reactions that keep organisms alive.

NEANDERTHALS

Homo neanderthalensis, a species of humans that interbred with Denisovans, *Homo sapiens* and possibly other human species. They lived in Asia and Europe from about 400,000 years ago until they became extinct about 40,000 years ago. They are genetically our closest extinct relatives but probably not directly in our lineage. Most human genomes today carry a small amount their DNA. Competition from *Homo sapiens* migrating out of Africa probably played a role in their extinction, although other factors also may have been important.

NUCLEOTIDES

The chemical building blocks of DNA [consisting of adenine, thymine, cytosine and guanine] and RNA [consisting of adenine, thymine, cytosine and uracil]. There are many other types of nucleotides in nature other than those in DNA and RNA.

ORGANELLES

Structures within cells, often having their own membranes, that perform specific life functions such as providing energy, building proteins and eliminating waste.

ORGANIC

Carbon-containing compounds that are fundamental for life as we know it on Earth.

OSMOTIC

The force or pressure causing the movement of water across membranes from low concentrations to high concentrations.

POLYMER

A chain of many small, repeating units. DNA and RNA are polymers, as are proteins and sugars.

PLURIPOTENTIAL STEM CELL

In biology, cells that have the potential to develop into almost any mature natural cell type. These cells are a key component of natural embryonic development. Scientists have learned how to manipulate them so they can become specific cell types for use in synthetic biology.

QUANTUM

A discrete physical quantity such as energy, momentum, charge or angular momentum that can only take on specific, fixed values—each step being one 'quantum.' In an atom, electrons exist at discreet energy 'levels' and can only move to a higher or lower level in quantum leaps. There is no intermediate state between energy levels in an atom.

QUANTUM MECHANICS

The fundamental physical theory that describes the behavior of matter and light at atomic and subatomic scales.

RIBOSOME

The essential component of all living cells responsible for converting amino acids into proteins. They are chemical manufacturing complexes residing within the fluids of all living cells. They consist of RNA plus certain proteins.

RNA

Ribonucleic acid. DNA contains the genetic blueprint of every living organism (with the possible exception of RNA viruses). RNA is the molecule that makes it all happen. Unlike DNA, it consists of a single rather than double strand and exists in many different forms in a typical organism. These include messenger RNA (mRNA), which communicates the DNA sequence to the ribosomes; transfer RNA (tRNA), which selects the appropriate amino acid for incorporation into a protein; and several types of regulatory RNA, which are involved in epigenetic processes and many other functions. RNA is a catalyst for many cellular reactions. It is possibly the basis for the origination of all life on Earth.

SYMBIOTIC

A beneficial relationship between species.

REFERENCES

1. "Life" (Merriam-Webster), accessed March 13, 2026, https://www.merriam-webster.com/dictionary/life.

2. https://www.sciencenewstoday.org/could-life-exist-without-water.

3. Sylvia Berryman, "Democritus" (*Stanford Encyclopedia of Philosophy*), last modified 7 January 2023, https://plato.stanford.edu/entries/democritus.

4. Gordon Campbel, "Empedocles (c. 492–432 B.C.E.)" (*Internet Encyclopedia of Philosophy*), accessed 13 March 2026, https://iep.utm.edu/empedocles.

5. Plato's Timaeus: https://www.gutenberg.org/ebooks/1572?utm_source=copilot.com

6. Aristotle's view of the soul: https://classics.mit.edu/Aristotle/soul.html.

7. Pasteur's view of vital action: https://www.bing.com/search?q=louis+pasteur+view+of+vital+action+theory&qs=GS&pq=louis+pasteur+view+of+vital+action&sc=12-34&cvid=38B83FF46D2F42C38B8E06CB541A86E9&FORM=QBLH&sp=1&ghc=1&lq=0

8. René Descartes' mind-body dulaism: https://iep.utm.edu/descartes-mind-body-distinction-dualism/?copilot_analytics_metadata=eyJldmVudEluZm9fY2xpY2tTb3VyY2UiOiJjaXRhdGlvbkxpbmsiLCJldmVudEluZm9fY29udmVyc2F0aW9uSWQiOiJ3V2lhdDVhTDdZS0pwZXVIcHlYOWkiLCJldmVudEluZm9fY2xpY2tEZXN0aW5hdGlvbiI6Imh0dH

9. Erwin Schrödinger, *What Is Life?* (Cambridge: Cambridge University Press, 1944).

10. Gyula Palyi, Claudia Zucchi and Luciano Caglioti (eds), *Fundamentals of Life* (Paris: Elsevier, 2002), *2*.

11. Gyula Palyi, Claudia Zucchi and Luciano Caglioti (eds), F*undamentals of Life* (Paris: Elsevier, 2002), *50*.

12. Jean Gayon, Christophe Malaterre, Michel Morange, Florence Raulin-Cerceau and Stéphane Tirard, "Defining Life: Conference Proceedings," *Origins of Life and Evolution of Biospheres* 40 (2010): 119–244.

13. Astrobiology at NASA, "About Life Detection," accessed June 12, 2026, https://astrobiology.nasa.gov/research/life-detection/about/

14. John D. Rummel, Lynn Harper and Dale Andersen, "Exobiology: The NASA Program" (NASA), accessed March 13, 2026, https://ntrs.nasa.gov/citations/19930009372.

15. Jean Gayon, Christophe Malaterre, Michel Morange, Florence Raulin-Cerceau and Stephane Tirard, "Defining Life: Conference Proceedings," *Origins of Life and Evolution of Biospheres* 40 (2010): 119–244, *236*.

16. Thomas R. Cech, *The Catalyst* (New York: W.W. Norton & Company, 2024).

17. Reed Bender, Karina Kofman, Blaise Agüera y Arcas and Michael Levin, "What Lives? A Meta-analysis of Diverse Opinions of the Definition of Life" (ArXiv), accessed June 15, 2026, https://arxiv.org/pdf/2505.15849.

18. Daniel Koshland, "The Seven Pillars of Life," *Science* 295 (2002): 2215–2216.

19. Y. N. Zhuravlev and V. A. Avetisov, "The Definition of Life in the Context of Its Origin," *Biogeosciences* 3 (2006): 281–291, 282.

20. Rebecca Skloot, *The Immortal Life of Henrietta Lacks* (New York: Crown, 2010).

21. Brett J. Kagan et al, "*In Vitro* Neurons Learn and Exhibit Sentience When Embodied in a Simulated Game-World," *Neuron* 110, no. 23 (2022): 3952–3968.

22. Douglas Blackiston et al, "A Cellular Platform for the Development of Synthetic Living Machines," *Science Robotics* 6, no. 52 (2021), accessed 2 April 2026, https://doi.org/10.1126/scirobotics.abf1571.

23. The Miller-Urey experiment is the most famous attempt described in this Wikipedia article: https://en.wikipedia.org/wiki/Miller–Urey_experiment?utm_source=copilot.com

24. Daniel G. Gibson, John I. Glass, Carole Lartigue and Vladimir N. Noskov, "Creation of a Bacterial Cell Controlled by a Chemically Synthesized Genome," *Science* 329, no. 5987 (2010): 52–56.

25. NIST, "Scientists Develop Cell with Synthetic Genome that Grows and Divides Normally," March 29, 2021, https://www.nist.gov/news-events/news/2021/03/scientists-develop-cell-synthetic-genome-grows-and-divides-normally.

ABOUT THE AUTHOR

Don Simborg, MD, is a physician, medical informaticist and author with over five decades of experience in healthcare technology. He received his MD degree from the Johns Hopkins School of Medicine and is board certified in Internal Medicine.

Dr Simborg's deep interest in life sciences, human evolution, artificial intelligence and genetic engineering has led him to explore the profound questions about our specie's past, present and future. He is the author of *What Comes After Homo Sapiens* and *The Fourth Great Transformation*. His books synthesize complex scientific concepts into accessible narratives for curious readers.

Dr Simborg pioneered the development of electronic health records and founded HL7, the preeminent standards organization for medical communications used worldwide. He has served on the Computer Science and Telecommunications Board of the National Research Council, National Academy of Sciences. His career has spanned clinical practice, academic research and entrepreneurship in the health information technology field.

Throughout his distinguished career, Dr Simborg has been at the forefront of applying technology to medicine. He has witnessed firsthand the transformative power of innovation in healthcare and has dedicated much of his later work to understanding how emerging technologies will shape not just medicine, but the very nature of humanity itself.

His unique perspective as both a practicing physician and a technology pioneer gives him rare insight into the intersection of biology and technology – a central theme in all his books. Dr Simborg brings a scientifically rigorous yet accessible approach to some of the most profound questions facing human evolutionary and medical science.